Module 2
Foundations in chemi

Atoms and reactions

Atomic structure and isotopes

For this topic you need to:

- know the relative masses and charges of protons, neutrons and electrons
- be able to quote the numbers of these particles in an atom or ion when given the atomic number and the mass number
- learn the definitions of the terms *relative atomic mass* and *atomic mass*
- be able to calculate the relative atomic mass of an element from the relative abundances of its isotopes
- understand how a mass spectrogram can be used to provide this information
- understand the terms *relative molecular mass* and *relative formula mass*

1 a Complete the following table. (AO1) 2 marks

Particle	Relative charge	Relative mass
Proton (p)		
Neutron (n)		
Electron (e)		

b Give the number of protons, neutrons and electrons for each of the atoms or ions in the table below. (AO1) 5 marks

Atom or ion	$^{14}_{7}N^{3-}$	$^{40}_{20}Ca^{2+}$	$^{39}_{19}K$	$^{79}_{35}Br^{-}$	$^{81}_{35}Br^{-}$
Number of protons					
Number of electrons					
Number of neutrons					

2 a Define the term *relative atomic mass*. (AO1) 2 marks

....

....

....

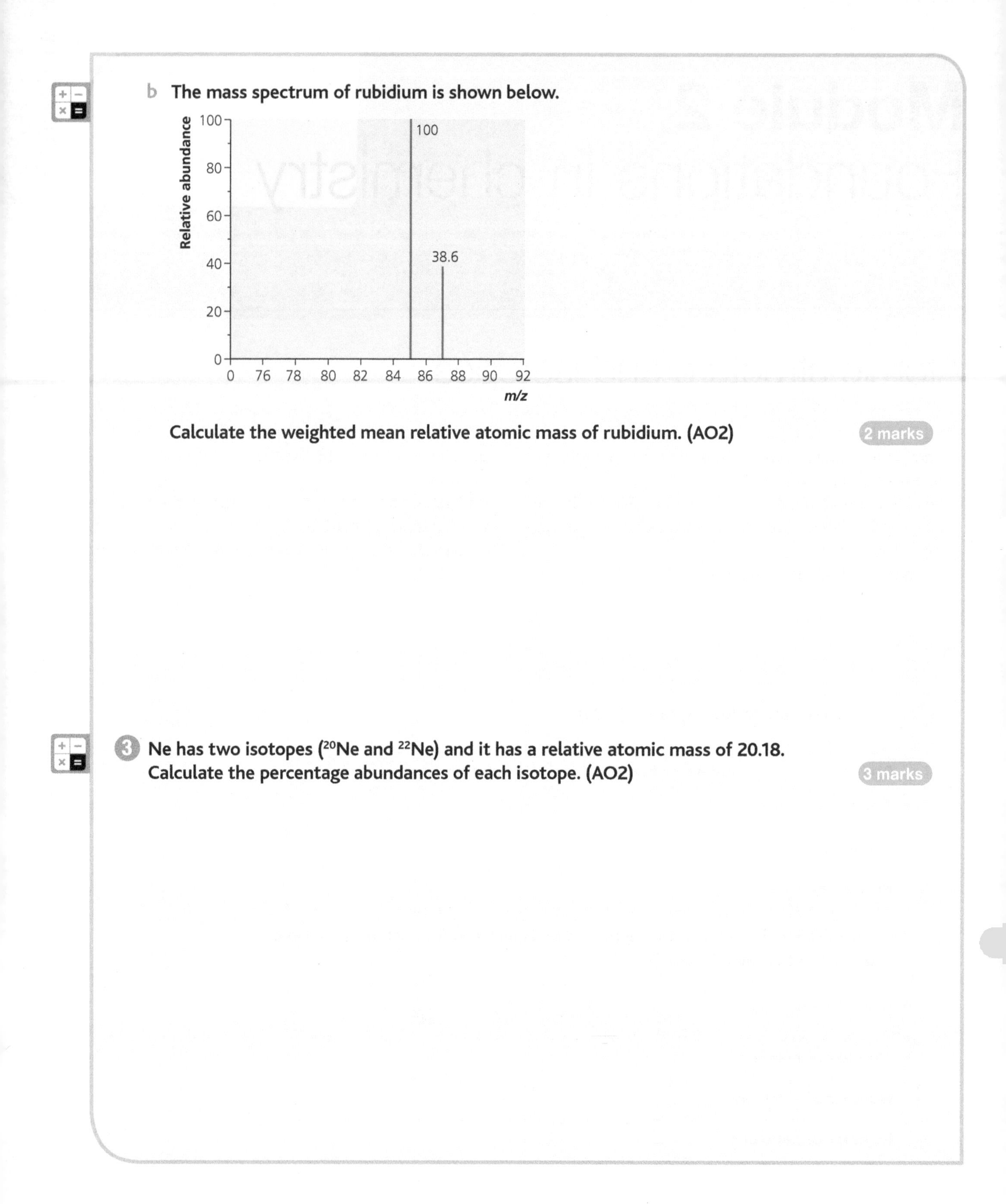

b The mass spectrum of rubidium is shown below.

Calculate the weighted mean relative atomic mass of rubidium. (AO2) 2 marks

3 Ne has two isotopes (^{20}Ne and ^{22}Ne) and it has a relative atomic mass of 20.18. Calculate the percentage abundances of each isotope. (AO2) 3 marks

Compounds, formulae and equations

Formulae and equations

Deriving the correct formulae for compounds and constructing balanced equations are basic requirements for the study of chemistry. The OCR specification details which ions you need to know, and making errors with these is a serious failing. Ionic equations include only those atoms, molecules or ions that have changed their state as a result of a reaction. State symbols are therefore always required for an ionic equation and are normally expected for other equations.

OCR

AS/A LEVEL YEAR 1

WORKBOOK

Chemistry A

Foundations in chemistry
Periodic table

John Older and Mike Smith

Contents

WORKBOOK

1 **This workbook will help you** to prepare for the following exams:

- OCR Chemistry A-level Paper 1: the exam is 2 hours 15 minutes long, worth 100 marks and 37% of your A-level.
- OCR Chemistry A-level Paper 2: the exam is 2 hours 15 minutes long, worth 100 marks and 37% of your A-level.
- OCR Chemistry A-level Paper 3: the exam is 1 hour 30 minutes long, worth 70 marks and 26% of your A-level.
- OCR Chemistry AS Paper 1: the exam is 1 hour 30 minutes long, worth 70 marks and 50% of your AS.
- OCR Chemistry AS Paper 2: the exam is 1 hour 30 minutes long, worth 70 marks and 50% of your AS.

2 **For each topic** there are:

- stimulus materials, including key terms and concepts
- short-answer questions
- long-answer questions
- questions that test your mathematical skills
- space for you to write

3 **Answering the questions** will help you to build your skills and meet the assessment objectives AO1 (knowledge and understanding), AO2 (application) and AO3 (analysis, interpretation and evaluation).

4 **You still need to** read your textbook and refer to your revision guides and lesson notes.

5 **Marks available** are indicated for all questions so that you can gauge the level of detail required in your answers.

6 **Timings** are given for the exam-style questions to make your practice as realistic as possible.

7 **Answers** are available at: www.hoddereducation.co.uk/workbookanswers

4 Rubidium (Rb) has an atomic number of 37, strontium (Sr) has an atomic number of 38 and selenium (Se) has an atomic number of 34.

Give the formulae of each of the following compounds. (AO2) 4 marks

Rubidium iodide

....

Rubidium carbonate

....

Strontium selenide

....

Rubidium selenide

....

5 The element cobalt (Co) is a transition element with an atomic number of 27. It has only one stable isotope with a mass number of 59. Cobalt forms 2+ ions in many of its compounds, but it can also form compounds containing the Co^{3+} ion.

a Give the numbers of protons, neutrons and electrons in an atom of cobalt. (AO1) 1 mark

....

b Give the formulae of cobalt(II) nitrate and cobalt(III) sulfate. (AO2) 2 marks

Cobalt(II) nitrate

....

Cobalt(III) sulfate

....

c Cobalt(II) carbonate reacts with dilute hydrochloric acid to form cobalt(II) chloride. Give an equation for this reaction. (AO2) 1 mark

....

d Cobalt also reacts slowly with aqueous silver nitrate to form silver metal and aqueous cobalt(II) nitrate. (AO2)

i Write an equation for this reaction. 1 mark

....

ii Write an ionic equation for this reaction. 1 mark

....

e Cobalt can also exist as a radioactive isotope (^{60}Co), which over time breaks down to form an isotope of nickel (^{60}Ni). During this transformation a stream of electrons is emitted from the atoms. (AO2, AO3)

i How does an atom of ^{60}Ni differ from an atom of ^{58}Ni? 1 mark

....

ii Suggest, in terms of the particles present in an atom, what changes have occurred when ^{60}Co breaks down into ^{60}Ni. 2 marks

....

....

....

....

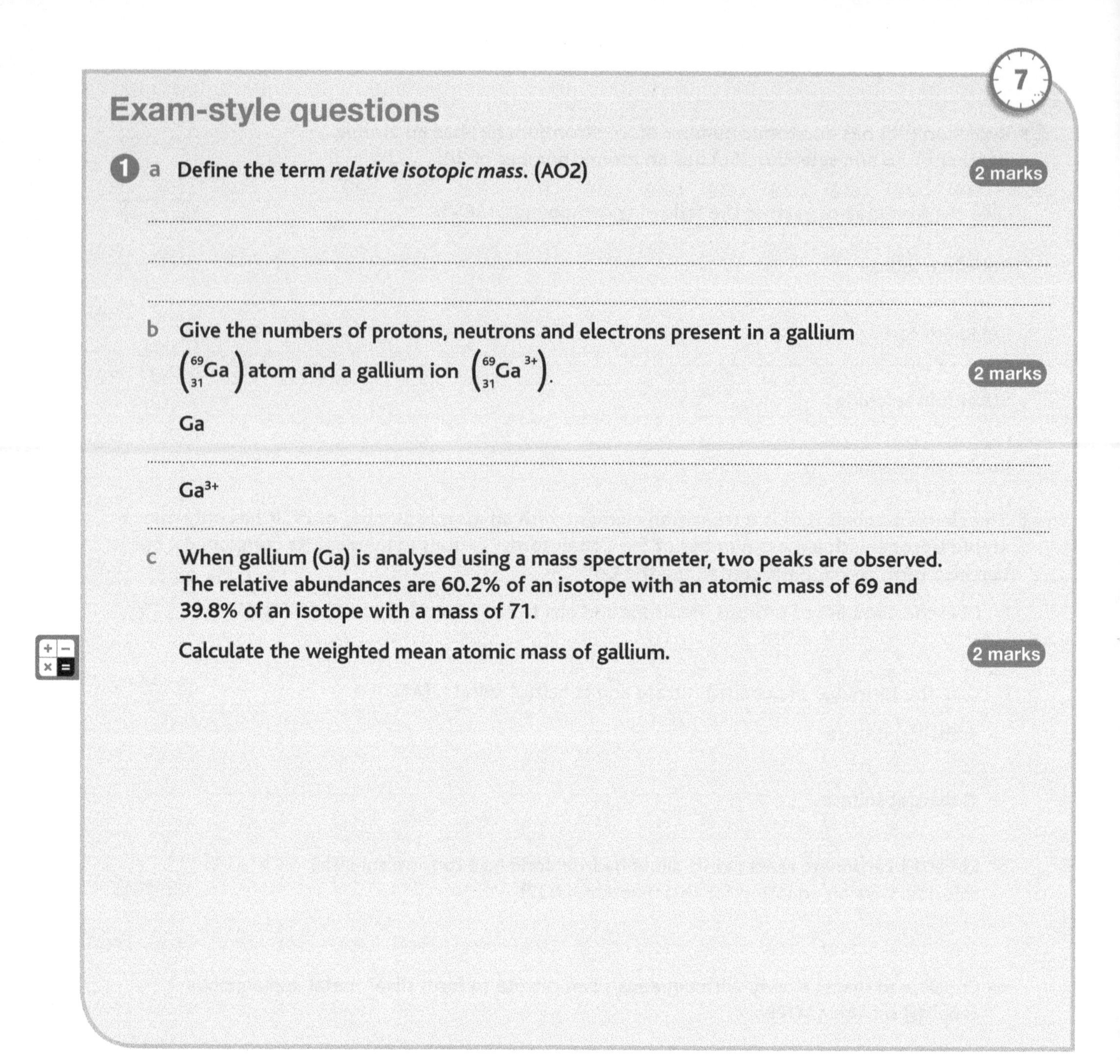

Exam-style questions

7

1 a **Define the term *relative isotopic mass*. (AO2)** 2 marks

....

....

....

b **Give the numbers of protons, neutrons and electrons present in a gallium ($^{69}_{31}Ga$) atom and a gallium ion ($^{69}_{31}Ga^{3+}$).** 2 marks

Ga

....

Ga^{3+}

....

c **When gallium (Ga) is analysed using a mass spectrometer, two peaks are observed. The relative abundances are 60.2% of an isotope with an atomic mass of 69 and 39.8% of an isotope with a mass of 71.**

Calculate the weighted mean atomic mass of gallium. 2 marks

Amount of substance

The mole

You need to be able to define some important terms, including the *mole*, the *Avogadro constant*, *molar mass* and the *molar gas constant*. The Avogadro constant is given on the Data Sheet in the exam and you must be able to use this to calculate the number of particles in a given mass of a substance.

1 a **Explain what is meant by:**

the Avogadro constant (AO1) 1 mark

....

molar mass (AO1) 2 marks

....

....

b i **Calculate how many atoms are present in 10.0 g of aluminium (AO2)** 2 marks

ii **Calculate how many chloride ions are present in 5.0 g of sodium chloride. (AO2)** 2 marks

Empirical formulae

The empirical formula shows the ratio of the amounts in moles of each element present in a compound expressed in its simplest whole number form. It is usually deduced from an analysis of the masses of each element present in a compound. For example, the analysis of a substance such as ethene (C_2H_4) would show that for every 1 mole of carbon atoms there are 2 moles of hydrogen atoms. So the empirical formula of ethene is CH_2.

The key relationship for converting masses into moles is:

$$\text{amount in moles of atoms of the element} = \frac{\text{mass of the atoms of the element}}{\text{mass of 1 mole of the element}}$$

2 **5.00 g of tin is reacted with an excess of iodine dissolved in an organic solvent until the tin has been completely converted into tin iodide. The excess iodine solution is removed and it is found that 26.38 g of tin iodide solid remains.**

a **Describe in detail how you would separate the iodine solution from the tin iodide and obtain a pure sample ready to be weighed. (AO3)** 3 marks

....

....

....

....

....

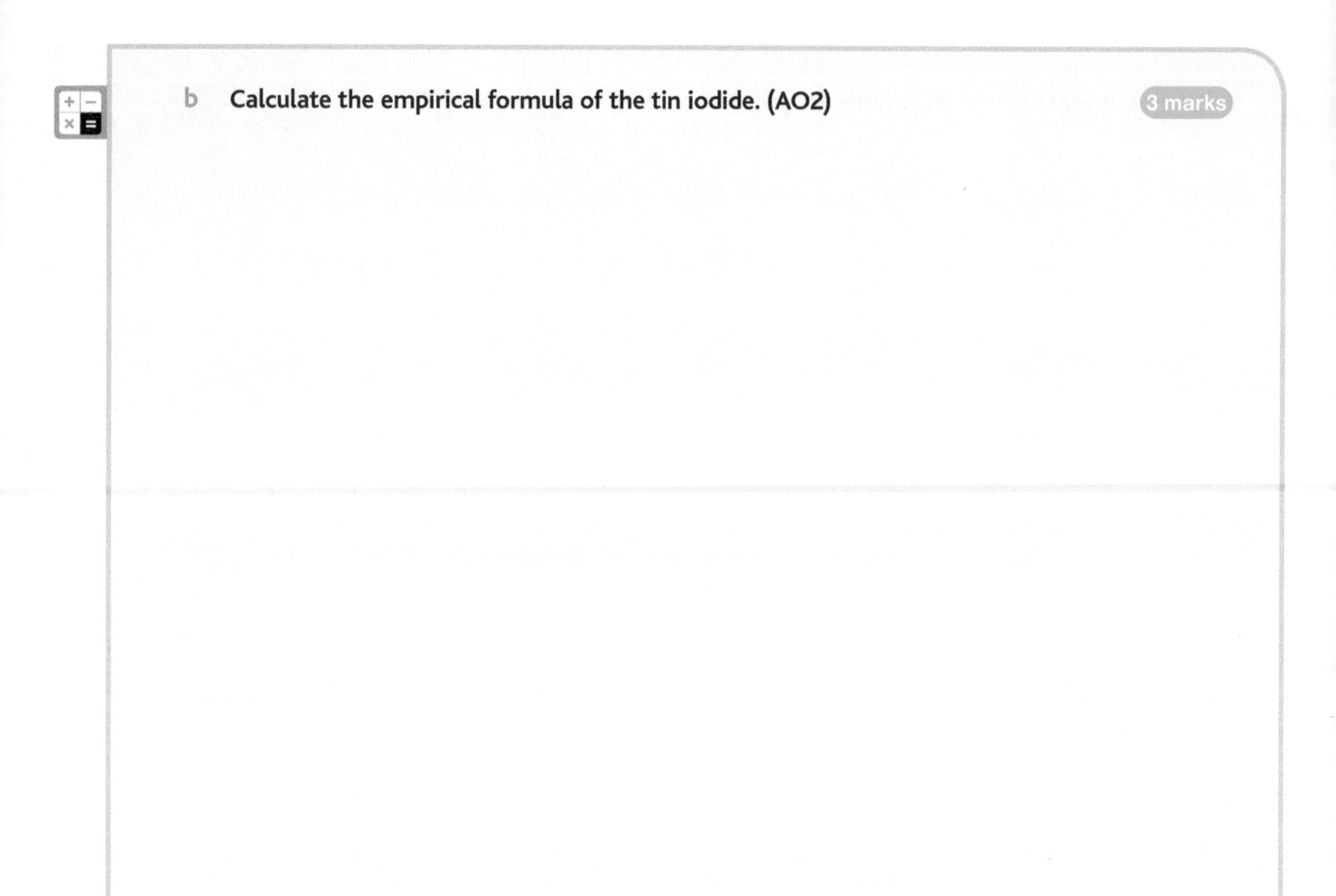

b **Calculate the empirical formula of the tin iodide. (AO2)** 3 marks

Formula of a hydrated salt

A value for the amount in moles of water of crystallisation in an inorganic crystal can be obtained by experiment. The experiment involves finding the mass of the crystal and then the mass of the anhydrous substance once the water has been driven off by heating. The masses of the anhydrous substance and of the water driven off are converted into moles by dividing them by the molecular mass of each compound.

3 Using the apparatus shown below, crystals of iron(II) sulfate are converted into anhydrous iron(II) sulfate.

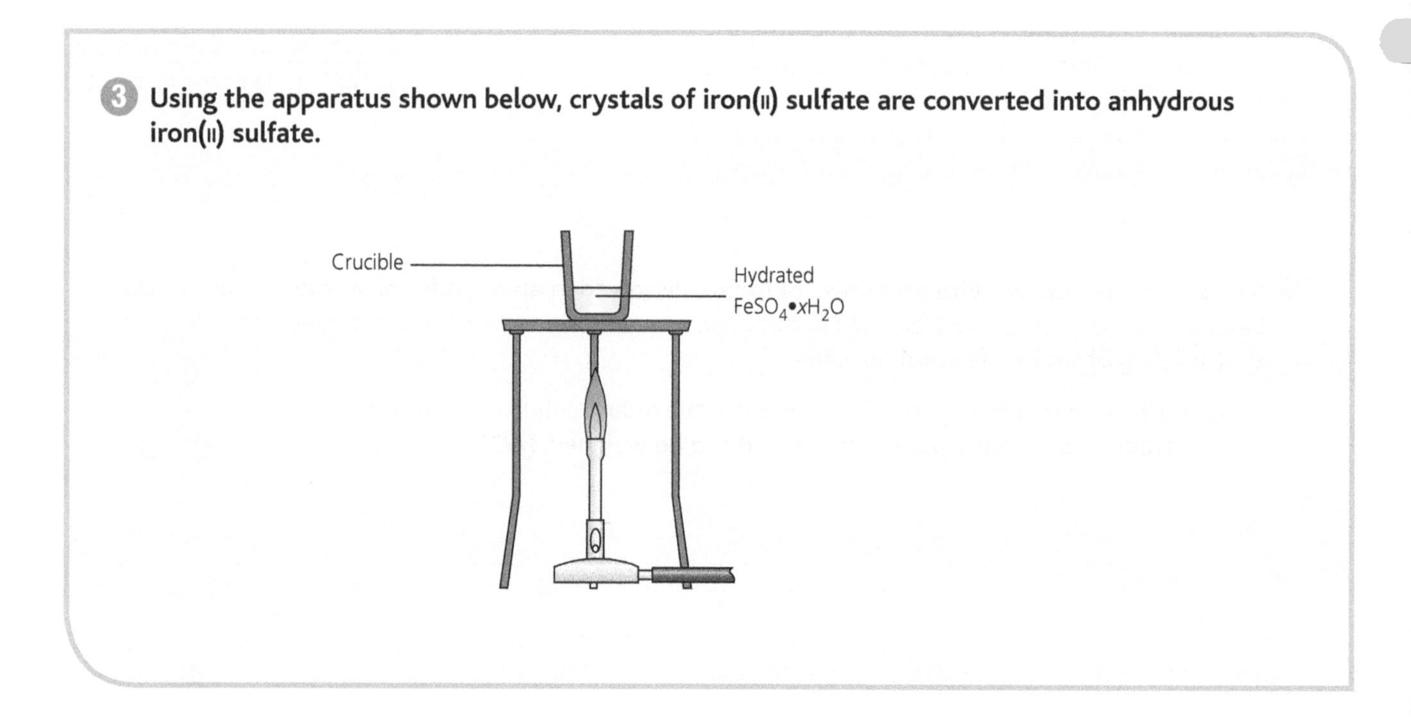

a **Use the following results from the experiment to determine the value of *x* in $FeSO_4 \cdot xH_2O$. (AO2)** 3 marks

mass of crucible = 20.23 g mass of $FeSO_4 \cdot xH_2O$ = 28.91 g mass of $FeSO_4$ formed = 24.98 g

b **In the experiment, how could you make sure that the measurement of the mass of the anhydrous solid was reliable? (AO3)** 1 mark

....

....

Calculations of reacting masses and gas volumes

In the laboratory, quantities are measured by taking their mass or volume, but chemical equations measure amounts in units called moles. To use the information that an equation provides, masses or volumes must first be converted into amounts in moles using the molar mass or the molar volume of substances. The key relationships are:

$$\text{amount in moles of a solid} = \frac{\text{mass of the solid}}{\text{mass of 1 mole of the solid}}$$

$$\text{amount in moles of a gas} = \frac{\text{volume of the gas in cm}^3}{\text{volume of 1 mole of gas}}$$

The volume of a gas changes with pressure or temperature, but a standard condition known as RTP is often used. This is the volume that 1 mol of the gas would occupy at room temperature (298 K) and standard pressure (100 kPa) and is taken to be 24.0 dm^3 or 24 000 cm^3.

You should have carried out a range of test-tube experiments that allow you to observe the reactions of acids. In questions of this kind you may also have to include a description of observations made as the reaction occurs.

4 **When 0.50 g of magnesium carbonate is added to an excess of dilute hydrochloric acid, a reaction takes place and 142 cm^3 of carbon dioxide is collected in a syringe at RTP.**

a **Describe what you would see when magnesium carbonate reacts with excess hydrochloric acid. (AO1)** 2 marks

....

....

b **Write the equation for the reaction between magnesium carbonate and hydrochloric acid. Include state symbols. (AO1, AO2)** 2 marks

....

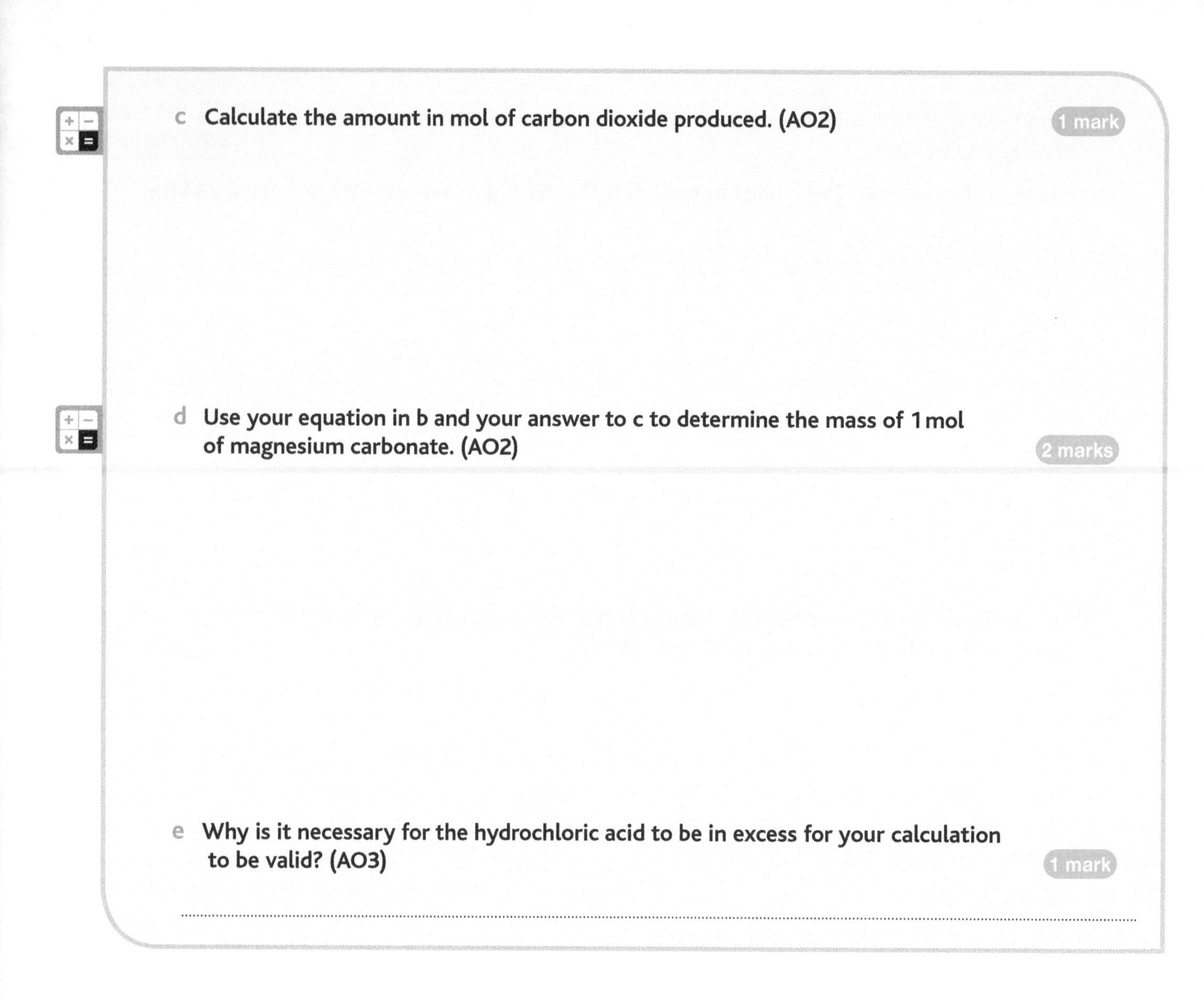

c **Calculate the amount in mol of carbon dioxide produced. (AO2)** 1 mark

d **Use your equation in b and your answer to c to determine the mass of 1 mol of magnesium carbonate. (AO2)** 2 marks

e **Why is it necessary for the hydrochloric acid to be in excess for your calculation to be valid? (AO3)** 1 mark

Volumes of gases from equations

Because 1 mol of gas has a fixed volume it is possible to deduce the volumes required for a reaction directly from an equation. The equation:

$$CH_4(g) + 2O_2(g) \rightarrow CO_2(g) + 2H_2O(g)$$

indicates that twice the volume of oxygen as methane is needed for a complete reaction. Carbon dioxide and water vapour are also produced in the proportions shown by the equation.

For example, 100 cm³ of methane requires 200 cm³ of oxygen for a complete reaction to produce 100 cm³ of carbon dioxide and 200 cm³ of water vapour.

This is straightforward, but it applies only to volumes of *gases*. If the equation had indicated the H_2O was produced as liquid water ($H_2O(l)$), then it would not be correct to conclude that its volume was 200 cm³.

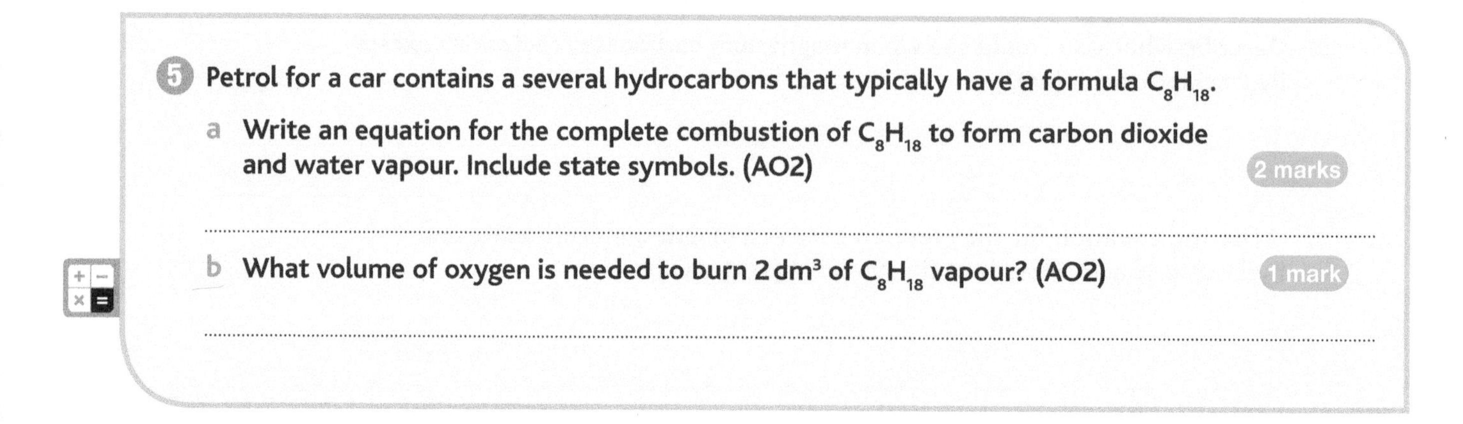

5 **Petrol for a car contains a several hydrocarbons that typically have a formula C_8H_{18}.**

a **Write an equation for the complete combustion of C_8H_{18} to form carbon dioxide and water vapour. Include state symbols. (AO2)** 2 marks

b **What volume of oxygen is needed to burn 2 dm³ of C_8H_{18} vapour? (AO2)** 1 mark c **What is the total volume of products obtained if 2 dm³ of C_8H_{18} vapour is burned to produce carbon dioxide and water vapour? (AO2)** 1 mark

....

d i **If the supply of oxygen is limited, some of the C_8H_{18} may react to form carbon monoxide instead of carbon dioxide. Write an equation to show the incomplete combustion of 1 mol of C_8H_{18} to produce only 6 mol of carbon dioxide with the rest of the carbon in the C_8H_{18} reacting to form carbon monoxide. (AO2)** 

....

ii **What difference would this make to the volume of the products obtained? (AO3)** 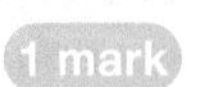

....

Concentrations of solutions

Solutions are often used in experiments and their concentrations can be expressed as the amount in moles of solid dissolved in 1 dm³ (1000 cm³) of solution. In this case:

amount in moles (n)
= volume of solution in dm³ (V)
× concentration of solution in mol dm⁻³

concentration in mol dm⁻³

$$= (\text{amount in moles in } V \text{ cm}^3 \text{ of solution}) \times \frac{1000}{V}$$

If the concentrations are given in g dm⁻³, you usually have to convert them into mol dm⁻³ during a calculation. This is done by dividing the mass in grams by the molar mass of the compound.

6 Calculate the concentration of a solution made by dissolving 5.0 g of sodium hydroxide to make 500 cm³ of solution. (AO2) 

7 How many moles are present in 25 cm³ of 0.1 mol dm⁻³ nitric acid? (AO2) 

8 What volume of water must be added to 50 cm³ of 2 mol dm⁻³ sodium nitrate solution to create a solution whose concentration is 0.1 mol dm⁻³? (AO2) 

Ideal gas equation

The ideal gas equation is:

$$pV = nRT$$

where p is the pressure, V is the volume, T is the temperature in kelvin, n is the amount of gas in mol and R is a constant. This equation shows how the volume of a gas is related to the amount in mol under differing conditions of temperature and pressure. When answering questions, remember that if you are using the value of R quoted on the Data Sheet as 8.314 J mol K^{-1}, then the pressure should be in kPa and the volume of the gas in dm^3. For these calculations you normally take 1 mol of gas as having a volume of 24.0 dm^3 at RTP.

It is often easier when answering questions to note that for a fixed amount in mol of gas:

$$\frac{p_1V_1}{T_1} = \frac{p_2V_2}{T_2}$$

where p_1, V_1 and T_1 are the pressure, volume and temperature under one set of conditions and p_2, V_2 and T_2 are the pressure, volume and temperature under a second set of conditions.

9 In an experiment a 1.40 g sample of calcium carbonate is heated until it has completely decomposed.

a **Write an equation for the decomposition. (AO1)** 1 mark

....

b **What is the volume of carbon dioxide collected as a result of the decomposition measured at 60°C and 100 kPa? (AO2)** 3 marks

c **If the experiment was repeated using 1.40 g of barium carbonate, would you expect the volume of carbon dioxide collected to be greater than the volume obtained from 1.40 g of calcium carbonate?**

Explain your answer. (AO2, AO3) 2 marks

....

....

....

....

Using other units

Exam questions may use units for mass other than grams. For example, you could be given a mass measured in metric tonnes, where 1 tonne = 10^6g. (This information is given on the Data Sheet that you are given in the exam.) There are several ways to approach such questions, but the easiest may be first to carry out any calculations in the usual way ignoring tonnes and using grams instead and then simply scale up the answer to tonnes. Remember that if 5.6g of a chemical reacts with 4.2dm^3 of a gas to form 4.8g of product, then 5.6×10^6g of the chemical reacts with 4.2×10^6dm^3 of the gas to form 4.8×10^6g of product. Or 5.6tonnes react with 4.2×10^6dm^3 of gas to form 4.8tonnes.

In all questions involving masses and volumes, be careful to check whether the answer must be given to a specific number of decimal places or significant figures. It is easy to forget to round the answer appropriately. However, note that if a calculation consists of several parts, you must not round the answer after each part. Instead, keep all the decimal places in your calculator and use them when you continue with the calculation.

10 **Iron can be extracted from its ore (Fe_2O_3) by reacting the ore with carbon monoxide (CO) at 1500°C in a furnace.**

a **Write an equation for the reaction to produce iron. Carbon dioxide is the other product. (AO1)**

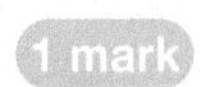

b **Calculate the volume of 1mol of carbon monoxide at 1500°C and standard pressure. Assume 1mol of gas at RTP has a volume of 24.0dm^3. Give your answer in dm^3 to 3 significant figures. (AO2)** 2 marks c Calculate the volume of carbon monoxide that is needed to react with the iron ore to produce 40 tonnes of iron at 1500°C. (AO2) 3 marks

d What volume of carbon dioxide is produced as a result of this reaction? (AO2) 1 mark

....

14

Exam-style questions

7

1 In the first row of the transition metals, cobalt ($_{27}Co$) with a relative atomic mass of 58.9 is placed before nickel ($_{28}Ni$), which has a relative atomic mass of 58.7.

a Give the electron configuration for $_{28}Ni$. 1 mark

$1s^2$ ....

b Usually as the atomic number increases, the relative atomic mass also increases. Explain why this is not true for cobalt and nickel. 3 marks

....

....

....

....

....

....

c When 2.00 g of nickel is reacted with bromine, 7.44 g of nickel bromide is formed. Calculate the empirical formula of nickel bromide. 2 marks

d i When aqueous sodium hydroxide is added to aqueous nickel nitrate, a precipitate of nickel hydroxide is obtained. Write an equation for the reaction. 1 mark

....

ii Give an ionic equation for this reaction. Include state symbols. 1 mark

....

2 Heating 2.51 g of a group 2 carbonate causes it to decompose and produce 483 cm^3 of carbon dioxide, measured at 80°C and standard pressure. 7

a Write an equation for the decomposition of the group 2 carbonate. Use M as the symbol for the metal. 1 mark

....

b Calculate the volume of the 483 cm^3 of the carbon dioxide if it had been measured at room temperature and pressure (RTP). 2 marks

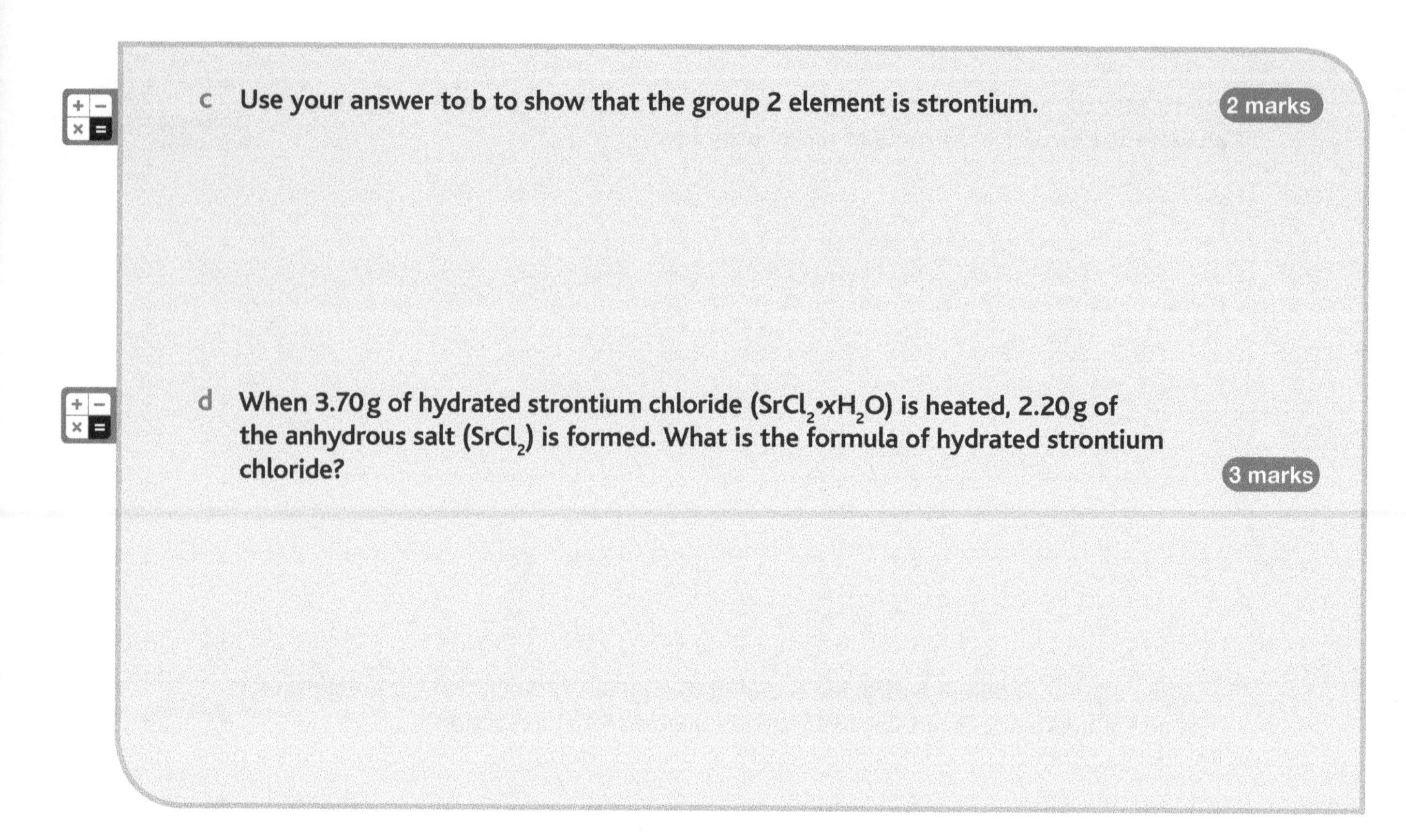

Acids

Acids, bases, alkalis and neutralisation

All acids release hydrogen ions in aqueous solution, although some are strong and fully ionised, while others are weak and not fully dissociated. Alkalis are soluble bases and release hydroxide ions in aqueous solution. Acids undergo a range of reactions in which the acid is neutralised to form a salt.

1 a **What is meant by a 'concentrated aqueous solution of a weak acid'? (AO1)** 2 marks

....

....

....

b i **Give an example of a weak acid and provide an equation to show its dissociation in water. (AO1, AO2)** 2 marks

....

....

ii **Write an equation to illustrate the reaction of an aqueous solution of this weak acid with aqueous sodium hydroxide. (AO1, AO2)** 2 marks

....

2 a **What is an alkali? (AO1)** 1 mark

....

b **Write an equation to show why ammonia is considered to be an alkali. (AO2)** 1 mark

....

c i Describe what you would see when aqueous sodium carbonate is neutralised by dilute nitric acid. (AO1) 1 mark

....

....

....

ii Write an equation for this reaction. (AO2) 1 mark

....

iii Write an ionic equation for the neutralisation of nitric acid by potassium hydroxide. (AO1) 1 mark

....

Acid–base titrations

A titration carried out carefully can be very accurate. You may be examined on the practical techniques needed to obtain reliable results. Problems are likely to be set based on the results obtained from a titration. This involves using the titre to determine the concentration of one of the solutions where the concentration of the other is known. Although you will have done this as a practical exercise, you must be ready to answer questions on it in the written exam.

3 a Describe the procedure for using a pipette to provide an accurate volume of a solution. (AO1) 3 marks

....

....

....

b A reading on a burette has a maximum error of $0.05\,cm^3$. What is the percentage error of a volume of $23.80\,cm^3$ measured using this burette? (AO3) 2 marks

c A student prepares a solution of sodium carbonate in a volumetric flask by dissolving some crystals in distilled water. However, instead of filling the flask correctly, the student uses too small a volume of water. Samples of this solution are then used in a titration with sulfuric acid.

Explain what effect this has on the calculated concentration of the sulfuric acid obtained from the titration. (AO2, AO3) 3 marks

....

....

....

....

....

4 25.0 cm^3 of a 0.25 mol dm^{-3} solution of sodium carbonate is titrated against hydrochloric acid. 18.5 cm^3 of the acid is required to reach the end-point.

a Write a balanced equation for the reaction that takes place. (AO1) 1 mark

....

b Calculate:

i the amount in mol in 25.0 cm^3 of the sodium carbonate solution (AO2) 1 mark

ii the amount in mol of hydrochloric acid that reacts with the sodium carbonate (AO2) 1 mark

iii the concentration of the hydrochloric acid in mol dm^{-3} (AO2) 1 mark

Exam-style questions

16

1 Below is a table of burette readings obtained from a titration of 0.100 mol dm^{-3} hydrochloric acid against 25.00 cm^3 samples of aqueous sodium hydroxide. The burette has an accuracy of 0.05 cm^3.

Titration number	Rough	1	2	3
Final volume/cm^3	23.00	22.32	22.60	44.95
Initial volume/cm^3	0	0	0	22.45
Volume used/cm^3	23.00	22.32	22.60	22.5

a The table above contains a number of errors in the recording of the results. Correct these errors using the table below. 3 marks

Titration number	Rough	1	2	3
Final volume/cm^3				
Initial volume/cm^3				
Volume used/cm^3				

b Use the results to give a value for the average titre to 2 decimal places. (AO3) 

c Use your answer to b to calculate the concentration of the aqueous sodium hydroxide. Give your answer to 3 significant figures. (AO2) 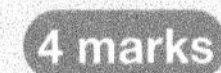

2 A student carries out a series of eight experiments. Volumes of 2.00 mol dm^{-3} hydrochloric acid are first added to a boiling tube using a burette. Then the same mass of marble (calcium carbonate) is added to each tube. In each experiment the volume of carbon dioxide given off is collected in a 250 cm^3 measuring cylinder inverted in water. 

The results obtained from each experiment are shown in the table below.

Experiment number	1	2	3	4	5	6	7	8
Volume of 2.00 mol dm^{-3} hydrochloric acid added/cm^3	1.0	2.0	3.0	4.0	6.0	8.0	10.0	12.0
Total volume of carbon dioxide collected/cm^3	27	45	66	91	115	117	114	116

a Write an equation for the reaction of marble and hydrochloric acid. Include state symbols. (AO1) 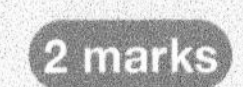

....

b The student notices that the reaction in experiments 5, 6, 7 and 8 has finished. What two observations would allow the student to observe that the reaction was complete in experiments 5, 6, 7 and 8? (AO3) 

....

....

c i On a separate sheet of graph paper, use the results to plot a graph of volume of carbon dioxide collected (*y*-axis) for volume of hydrochloric acid added (*x*-axis). (AO2) 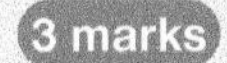

ii Draw best-fit lines on your graph to determine the volume of 2.00 $mol\,dm^{-3}$ hydrochloric acid that exactly reacts with the mass of marble used. (AO2) **2 marks**

Volume of hydrochloric acid to 1 decimal place = ...

d Use your answer to c ii to determine the mass of marble that this result suggests was used in each experiment. (AO2) **3 marks**

e What will be the concentration of the hydrochloric acid in the boiling tube once 8.0 cm^3 has been added? (AO2) **2 marks**

f The results of the experiment are not reliable. One source of error is the measuring cylinder used to collect the carbon dioxide. If this measuring cylinder has a maximum error of 2 cm^3, what is the percentage error in the first reading taken? (AO2) **1 mark**

g Apart from the error in the reading of the measuring cylinder, what other unavoidable problem that would affect the results might occur while collecting the carbon dioxide over water? (AO3) **1 mark**

....

Redox

You must be familiar with the rules for assigning oxidation numbers to elements in compounds or ions. You should be able to analyse a reaction to identify oxidation as the loss of electrons and reduction as the gain of electrons. This may be done by tracing the movement of electrons or using oxidation numbers. You need to know that the reactions of reactive metals with acids to form salts are examples of redox reactions and be able to apply these principles to other reactions where suitable information is given in a question.

1 a **Describe what you would see when zinc is reacted with dilute sulfuric acid. (AO1)** 2 marks

....

....

....

b **Write an equation for this reaction and use it to explain why it is considered to be a redox reaction. (AO2)** 3 marks

....

....

....

....

....

2 **When chlorine is bubbled into aqueous iron(II) chloride, the iron(II) chloride forms iron(III) chloride, but no other product is obtained.**

a **Write an equation for this reaction. (AO3)** 1 mark

....

b **Explain which ion or molecule has been oxidised and which has been reduced in this reaction. (AO2)** 2 marks

....

....

....

....

....

3 **One mole of copper reacts with 4 mol of concentrated nitric acid to make a salt containing Cu^{2+} ions and the gas nitrogen(IV) oxide is given off. Water is also formed.**

a **What is meant by a salt? (AO1)** 1 mark

....

....

b **Use the information given to provide an equation for this reaction. (AO3)** 1 mark

....

c Using oxidation numbers explain why this reaction is a redox reaction and state what has been oxidised and what has been reduced. (AO2) 3 marks

8

Exam-style questions

1 Bromine is a member of the halogen group of elements. It occurs naturally as two isotopes, one of which has a mass number of 79. It has a relative atomic mass of 79.9.

a State the numbers of protons, neutrons and electrons in a ^{79}Br atom. 1 mark

b The other isotope of bromine has two more neutrons in its nucleus. How does its percentage abundance compare with that of ^{79}Br? Explain your answer. 2 marks

c 25 cm^3 of 0.10 $mol\,dm^{-3}$ potassium bromide reacts with 50 cm^3 of 0.10 $mol\,dm^{-3}$ aqueous silver nitrate to form a precipitate of silver bromide.

i Give an ionic equation for this reaction. Include state symbols. 1 mark

ii Calculate the mass of silver bromide that is formed. 4 marks d Concentrated sulfuric acid reacts with sodium bromide according to the equation shown below:

$$2NaBr(s) + 2H_2SO_4(l) \rightarrow NaHSO_4(s) + Br_2(g) + SO_2(g) + 2H_2O(l)$$

Determine which element has been oxidised and which reduced by giving the change in oxidation number for each of the elements. 2 marks

Element that has been oxidised ....

Oxidation number change: from .... to ....

Element that has been reduced ....

Oxidation number change: from .... to ....

Electrons, bonding and structure

Electron structure

You need to know the orbital structures of the first 36 elements and understand the filling sequence of the orbitals, including the initial half-filling of orbitals where there is more than one of the same type (e.g. the three *p*-orbitals). You also need to know how the pattern of the first ionisation energies of the elements supports the pattern of energies assigned to orbitals.

Questions on this topic rarely expect more than straightforward knowledge.

1 Give the electron structure for each of the following. Show each *p*-orbital. (AO1) 5 marks

Li $1s^2$ ....

O $1s^2$ ....

Si $1s^2$ ....

S^{2-} $1s^2$ ....

Ca^{2+} $1s^2$ ....

2 a How does a 1*s*-orbital differ from a 2*s*-orbital? (AO1) 1 mark

....

....

b Draw a sketch of a 2*p*-orbital. (AO1) 1 mark

10

Exam-style questions

1 a **What is an electron orbital?** 1 mark

b **Give the electron configuration of:** 2 marks

i **neon**

ii **phosphorus**

c **The table below gives the values of the first ionisation energies of the elements neon (atomic number 10) to sulfur (atomic number 16).**

Element	Ne	Na	Mg	Al	Si	P	S
First ionisation energy/ $kJ\,mol^{-1}$	2081	496	738	578	789	1012	1000

Explain why the first ionisation energy of:

i **sodium is less than that of neon** 2 marks

ii **aluminium is less than that of magnesium** 2 marks

iii **silicon is more than that of aluminium** 2 marks

iv **sulfur is less than that of phosphorus** 2 marks

Bonding and structure

Ionic and covalent bonding

To explain the properties of different substances you need to understand the bonding in the substance and the type of structure associated with that bonding. The bonding is determined by the arrangement of electrons in the outer shells of the atoms. Atoms bond or combine either by losing or gaining electrons (ionic bonding) or by sharing electrons (covalent bonding). Ionic bonding produces a giant ionic lattice, whereas covalent bonding gives discrete individual molecules. You should be aware of the physical properties associated with each type of structure.

1 a Ionic bonding occurs when electrons are transferred from .... atoms to .... atoms. (AO2) 1 mark

b Explain what is meant by an *ionic bond*. (AO1) 1 mark

....

....

c Explain what is meant by a *covalent bond*. (AO1) 1 mark

....

....

d Explain what is meant by a *dative bond*. (AO1) 1 mark

....

....

2 Complete the table below to compare the properties of ionic, covalent and metallic substances. Explain your reasoning. (AO2) 6 marks

	Ionic	**Covalent**	**Metallic**
Melting point (high or low)	High / low*	High / low*	High / low*
	Explanation:	Explanation:	Explanation:
Conductivity (high or low)	High / low*	High / low*	High / low*
	Explanation:	Explanation:	Explanation:
	*delete as appropriate		

3 Draw 'dot-and-cross' diagrams of the following. Draw the outer electrons only. (AO2, AO3)

4 marks

Ammonia (NH_3)	Calcium oxide (CaO)
Phosphonium ion (PH_4^+)	**Carbonyl chloride ($COCl_2$)**

4 The data below give some properties of four substances, A, B, C and D. (AO2, AO3)

Compound	Solubility in water	Solubility in hexane	Electrical conductivity: Solid	Electrical conductivity: Liquid	Boiling point/°C
A	Good	Insoluble	Poor	Good	1465
B	Poor	Good	Poor	Poor	183
C	Poor	Poor	Poor	—	2355
D	Poor	Poor	Good	Good	2567

Explain how these data suggest different bonding and structures for the four compounds.

8 marks

Compound A

....

....

....

....

Compound B

....

....

....

Compound C

....

....

....

Compound D

....

....

....

Shapes of simple molecules and ions

The shape and the bond angles in covalent molecules can be determined by the electron-pair repulsion theory. You should be able to predict the shape of a molecule or ion with up to six pairs of electrons around the central atom. You should also be able to use Pauling electronegativity values to work out whether a bond is polar.

5 Explain the 'electron-pair repulsion theory'. (AO1) 3 marks

....

....

....

....

6 Draw diagrams to show the shape and the bond angles in these molecules. Explain why the molecule or ion has the shape you have drawn. (AO2, AO3) 9 marks

a **ammonia (NH_3)**

Shape is .... **and bond angle is** .... **because**

....

....

....

b **ammonium ion (NH_4^+)**

Shape is .... **and bond angle is** .... **because**

....

....

....

c **amide ion (NH_2^-)**

Shape is .... **and bond angle is** .... **because**

....

....

....

Electronegativity and bond polarity

The electrons in a covalent bond are not shared equally if the covalent bond is between atoms with different electronegativities. The difference in electronegativity can polarise the covalent bond and produce a permanent dipole. Sometimes the dipoles can cancel and result in a non-polar molecule.

7 **Explain what is meant by the term *electronegativity*. (AO1)** 1 mark

....

....

8 **Explain, with the aid of diagrams, why BCl_3 is a non-polar molecule and why PCl_3 is a polar molecule. (AO2)** 6 marks

....

....

....

9 **The Pauling electronegativity values of C, H, O and F are shown below.**

C	H	O	F
2.5	2.1	3.5	4.0

a **Add the dipoles to the bond highlighted in red in each of the following molecules. (AO2)** 5 marks

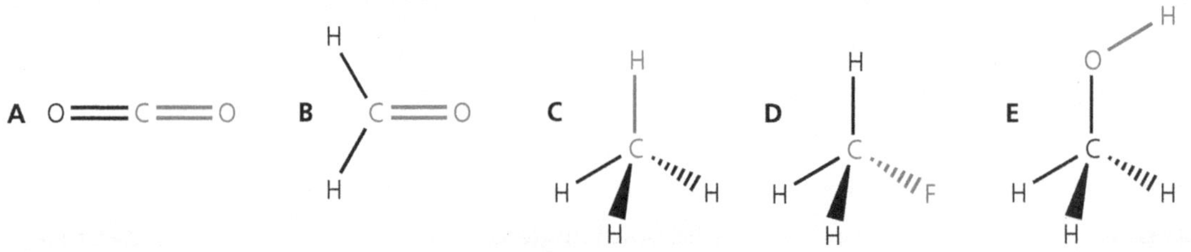

A O=C=O   B $H_2C=O$ C CH_4 D CH_3F E CH_3OH

b **Group the molecules A–E into one of the following two categories. (AO2, AO3)** 2 marks

Polar molecules ....

....

Non-polar molecules ....

....

Intermolecular forces

Intermolecular forces exist between molecules in substances. The intermolecular forces are a result of permanent and induced dipoles. In molecules that contain N, O or F bonded to an H atom, the intermolecular force is called a hydrogen bond.

10 Describe the intermolecular forces present in:

a hydrogen chloride gas (AO2) **2 marks**

b hydrogen gas (AO2) **2 marks**

c ammonia gas (AO2) **2 marks**

11 Give two anomalous properties of water and explain why these occur. (AO1) **4 marks**

22

Exam-style questions

1 This question compares the bonding in calcium, oxygen and calcium oxide. 12

a Calcium (Ca) is a metallic element. Explain, with the aid of a labelled diagram, what is meant by the term *metallic bonding*. **3 marks**

b Calcium reacts with oxygen to form calcium oxide.

i Write an equation, including state symbols, for this reaction. 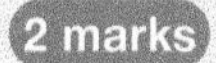

ii Draw dot-and-cross diagrams of oxygen and of calcium oxide. Show the outer electrons only. **3 marks**

Oxygen	Calcium oxide

iii Describe the type of bonding in oxygen. **1 mark**

....

....

c Compare and explain the electrical conductivity of calcium and of calcium oxide in the solid and the liquid states. **5 marks**

....

....

....

....

....

....

....

....

2 The boiling points of the hydrides of groups 14 and 16 are shown in Tables 1 and 2 below. (15)

Table 1 Hydrides of group 14

Compound	Boiling point/K	Number of electrons
CH_4	112	
SiH_4	161	
GeH_4	178	
SnH_4	221	

Table 2 Hydrides of group 16

Compound	Boiling point/K	Number of electrons
H_2O	373	
H_2S	213	
H_2Se	231	
H_2Te	270	

a Complete Tables 1 and 2 giving the total number of electrons in each compound. **4 marks**

b i Explain why the boiling point increases down group 14 from CH_4 to SnH_4. **2 marks**

....

....

....

....

ii The boiling points increase down group 16 from H_2S to H_2Te, but the boiling point of H_2O does not fit this trend. Explain, with the aid of a suitable diagram, why the boiling point of H_2O does not fit the trend shown by the other group 16 hydrides. **4 marks**

c i Draw dot-and-cross diagrams for H_2S and SiH_4. Show the outer electrons only. **2 marks**

ii Explain the difference in boiling point between H_2S and SiH_4. **4 marks**

Module 3 Periodic table

The periodic table

Periodicity

You are given a copy of the periodic table in the exam, so you do not need to learn its structure. However, you should know why the periodic table is arranged in the form that it has. There are several trends that are apparent in the periods of the table and these include the electron configuration of the elements, the first ionisation energies and the type of bonding. An understanding of these trends should allow you to predict and explain the variations in the melting points of the elements of periods 2 and 3 in terms of their structure and bonding. You should be able to recall that these essential features:

- The elements on the left-hand side of the periodic table have giant metallic lattice structures.
- Carbon in its various forms and silicon have giant covalent lattices.

1 a **How are the elements arranged in the periodic table? (AO1)** 1 mark

....

....

b **By considering their electron structures, explain why oxygen and sulfur are placed in the same group of the periodic table. (AO1)** 2 marks

....

....

....

c i **Explain what is meant by the *first ionisation energy of oxygen*. (AO1)** 2 marks

....

....

....

ii **Explain why the first ionisation energy of oxygen has a lower value than the first ionisation energy of fluorine. (AO2)** 2 marks

....

....

....

2 **The table below gives the successive ionisation energies of silicon.**

Ionisation	1st	2nd	3rd	4th	5th	6th
Energy/ kJ mol^{-1}	789	1577	3232	4356	16091	19785

a i Give an equation to represent the third ionisation energy of silicon. (AO1) 1 mark

ii After the third ionisation has taken place, what is the orbital structure of the ion formed? (AO2) 1 mark

iii Explain why the second ionisation energy of silicon is greater than the first ionisation energy. (AO2) 2 marks

iv How do the ionisation energies confirm that silicon is a member of group 14 of the periodic table? (AO2) 1 mark

b Describe the structure of silicon. (AO1) 2 marks

3 Explain how magnesium conducts electricity. (AO2) 1 mark

a Describe, with the aid of a diagram, the structure of graphene. Explain why graphene conducts electricity. (AO2) 3 marks

b How is the structure of graphene related to the structure of graphite? (AO2) 2 marks

9

Exam-style questions

1 **The table below gives the melting point of three elements in period 2 of the periodic table.**

Element	Lithium	Carbon (diamond)	Neon
Melting point/°C	180.5	3527	−250

a **Explain these melting points in terms of the structure and bonding of each element.** 6 marks

....

....

....

....

....

....

....

....

....

....

....

b **Complete the table below giving the type of lattice structure shown by the elements of the third period.** 2 marks

Element	Mg	Al	Si	P
Type of structure				

c **Draw a sketch to show the pattern of the first ionisation energies of magnesium, aluminium, silicon, phosphorus and sulfur. The ionisation energy of magnesium is indicated by a cross (×) on the graph below.** 3 marks

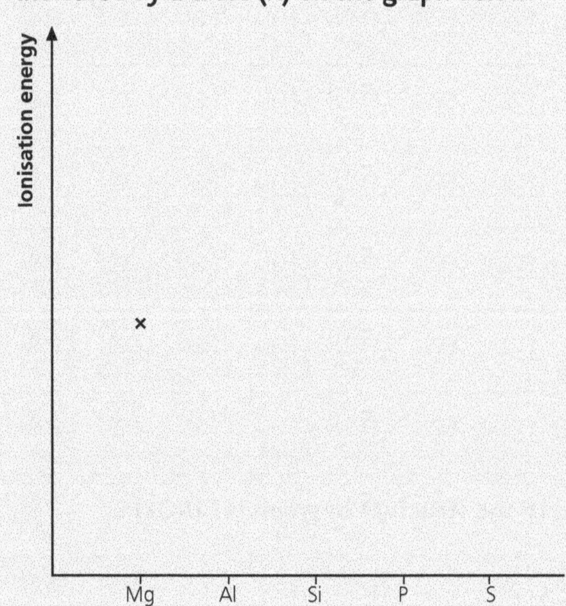

Group 2

Redox reactions and reactivity of group 2 metals

You should know the full electron configurations of the group 2 atoms and of the ions formed by the loss of the s^2 outer electrons when the group 2 atoms react. It follows that reactions of the group 2 metals are redox reactions, with the group 2 metal behaving as the reducing agent. The relative reactivities of the metals can be justified by the size of the first and second ionisation energies.

1 **Write the full electron configuration for:**

a $_{20}Ca$ **(AO1)** 1 mark

....

b $_{38}Sr^{2+}$ **(AO2)** 1 mark

....

2 a **Define the *first ionisation energy*. (AO1)** 2 marks

....

....

....

b **Write an equation to show:**

i **the first ionisation energy of Ba (AO2)** 2 marks

....

ii **the second ionisation energy of Be (AO2)** 2 marks

....

3 **The first, second and third ionisation energies for group 2 elements are shown in the table below.**

Ionisation energy/kJ mol^{-1}	Be	Mg	Ca	Sr	Ba
1st	900	738	590	550	503
2nd	1757	1451	1145	1064	965
3rd	14849	7733	4912	4210	3600

a **Explain the trend in first ionisation energy down the group from Be to Ba. (AO2)** 4 marks

....

....

....

....

....

b i **State the trend in second ionisation energy down the group from Be to Ba. (AO2)** 1 mark

....

ii Explain why, for each element, the second ionisation energy is always greater than the first ionisation energy. (AO2, AO3) 3 marks

....

....

....

....

....

iii Explain why, for each element, the third ionisation energy is always considerably greater than the second ionisation energy. (AO2, AO3) 4 marks

....

....

....

....

....

....

....

4 Explain the trend in reactivity down group 2. (AO2) 4 marks

....

....

....

....

....

....

....

5 Write equations for each of the following reactions:

a magnesium and oxygen (AO2) 1 mark

....

b strontium and water (AO2) 1 mark

....

c barium and hydrochloric acid (AO2) 1 mark

....

d barium and ethanoic acid (AO2, AO3) 1 mark

....

e barium and lactic acid ($CH_3CH(OH)COOH$) (AO3) 1 mark

....

6 Write ionic equations for each of the following:

a strontium and water (AO2) 1 mark

....

b barium and hydrochloric acid (AO2) 1 mark

....

c barium and ethanoic acid (AO2, AO3) 1 mark

....

d barium and lactic acid ($CH_3CH(OH)COOH$) (AO3) 1 mark

....

Reactions of group 2 compounds

You should know the reaction of water with group 2 oxides and understand the alkalinity and uses of some of the resulting group 2 compounds.

7 Write equations and ionic equations for each of the following reactions:

a barium oxide solid + water → ... (AO1, AO2) 3 marks

Equation

....

Ionic equation

....

b magnesium hydroxide solid suspension + sulfuric acid → ... (AO1, AO2) 3 marks

Equation

....

Ionic equation

....

c calcium carbonate solid + aqueous phosphoric acid (H_3PO_4) → ... (AO2, AO3) 3 marks

Equation

....

Ionic equation

....

The halogens (group 17)

Characteristic physical properties

Halogens exist as diatomic molecules and down group 17 their physical state changes from gas to liquid to solid.

1 a **Draw a dot-and-cross diagram of chlorine showing the outer electrons only. Explain why chlorine is diatomic. (AO1)** 2 marks

....

....

b **State the type of bonding in chlorine and explain whether or not chlorine is a polar molecule. (AO1, AO2)** 2 marks

....

....

2 **The boiling points of the halogens are shown in the table below.**

Halogen	Fluorine	Chlorine	Bromine	Iodine
Boiling point/°C	–188	–35	59	184*
Physical state at 25°C				

(* iodine sublimes when heated)

a **Complete the table above by writing the physical state of each element. (AO1)** 1 mark

b **Explain the trend in boiling points of the halogens. (AO1, AO2)** 3 marks

....

....

....

....

....

....

Redox reactions and reactivity of halogens and their compounds

You should know the outer-shell configuration of the halogens and be aware that they react by gaining one electron. You should also know that halogens are diatomic molecules, while halides are negatively charged ions.

In group 17 the trend is for reactivity to decrease down the group and you should be able to explain why. Reactions of halogens are redox reactions in which they usually behave as oxidising agents. You should know that a halogen displaces a heavier halide from one of its salts. In certain circumstances the halogen can be simultaneously oxidised and reduced. Chlorine is used in water treatment and in the manufacture of bleach.

3 Write the full electron configuration for:

a $_{35}Br$ **(AO1)** 1 mark

....

b $_{17}Cl^{-}$ **(AO2)** 1 mark

....

4 Explain the trend in reactivity down group 17. (AO2) 5 marks

....

....

....

....

....

5 a Complete the table below and indicate with a tick (✓) if a reaction occurs and a cross (✗) if a reaction does not take place. Some have been done for you. (AO2) 2 marks

	Fluoride (F^-)	Chloride (Cl^-)	Bromide (Br^-)	Iodide (I^-)
Fluorine (F_2)				
Chlorine (Cl_2)	✗		✓	✓
Bromine (Br_2)				
Iodine (I_2)				

b **Write an ionic equation for the reaction that would occur between chlorine and aqueous potassium bromide. (AO2)** 1 mark

....

c **Chlorine gas is bubbled into a test tube containing aqueous potassium iodide. When the reaction is complete, hexane ($C_6H_{14}(l)$) is then added to the test tube and shaken.**

i **Describe, and explain, your observations when chlorine gas is bubbled into the test tube containing aqueous potassium iodide. (AO1)** 1 mark

....

....

....

ii Describe your observations when hexane is then added and the mixture shaken. 1 mark

6 a Explain what is meant by the term *disproportionation*. (AO1) 2 marks

b The reaction of chlorine with water is used in water purification. (AO2)

i Write an equation, including state symbols, for the reaction between chlorine and water. (AO2) 2 marks

ii Write an ionic equation for the reaction between chlorine and water. (AO3) 2 marks

iii State the benefits and the risks associated with chlorination of water. (AO1) 2 marks

7 Chlorine can react with sodium hydroxide to form bleach. Write an equation for this reaction. State the necessary conditions. (AO2) 2 marks

Equation

Conditions

8 Hydrogen sulfide gas is bubbled into an aqueous solution of chlorine and a yellow precipitate of sulfur is observed. The resulting solution turns blue litmus red.

a Construct an equation, including state symbols, for this reaction. (AO2) 2 marks

b Identify the oxidising agent in this reaction. Justify your answer. (AO2, AO3) 2 marks

Characteristic reactions of halide ions

The halides of most metals are soluble in water, but silver halides are insoluble. Aqueous chlorides, bromides and iodides can therefore be detected by the addition of aqueous $Ag^{+}(aq)$ ions. The silver halide precipitates have differing solubility in ammonia.

9 **Complete the table below. In each reaction the aqueous solutions were mixed and the initial observation recorded. Dilute ammonia was added and the observation recorded. Concentrated ammonia was then added and the final observation recorded. (AO2)** 12 marks

		Observations		
Reaction	**Ionic equation**	**Initial**	**Dil. NH_3**	**Conc. NH_3**
$KCl(aq)$ + $AgNO_3(aq)$				
$CaCl_2(aq)$ + $AgNO_3(aq)$				
$NaBr(aq)$ + $AgNO_3(aq)$				
$BaI_2(aq)$ + $AgNO_3(aq)$				

Qualitative analysis

Tests for ions

You should know the simple test-tube reactions that can be used to test for the presence of:

- anions (carbonate, sulfate and halides)
- cations (ammonium)

and the sequence in which the tests should be carried out.

10 Describe the test for each of the following. (AO2)

a a carbonate 3 marks

Reagent

....

Equation

....

Observation(s)

....

b a bromide 3 marks

Reagent

....

Equation

....

Observation(s)

....

c a sulfate 3 marks

Reagent

....

Equation

....

Observation(s)

....

d State the sequence in which the three tests above should be carried out. Explain why. (AO2) 2 marks

....

....

....

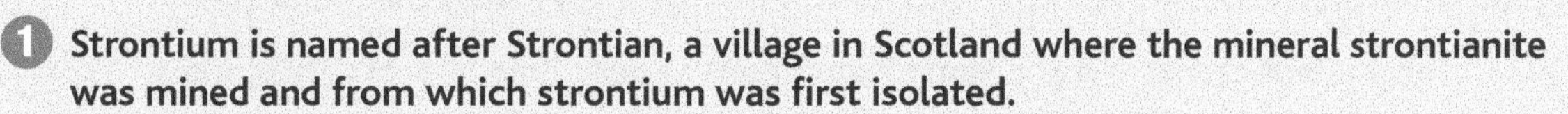

Exam-style questions

1 Strontium is named after Strontian, a village in Scotland where the mineral strontianite was mined and from which strontium was first isolated. (10)

a Strontium reacts more readily with water than calcium does.

i Write an equation, including state symbols, for the reaction of strontium with water.  2 marks

ii Write an ionic equation for the reaction of calcium with water. 1 mark

iii Explain why strontium reacts faster with water than calcium does. 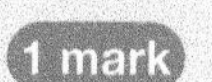 3 marks

b Strontium burns in the air to form both strontium oxide and strontium nitride.

i Write an equation for the reaction of strontium with oxygen.  1 mark

ii State the formula of strontium nitride. 1 mark

c Dilute hydrochloric acid is added to a solution of strontium carbonate to produce a solution A and a gas B. Solution A is divided into two parts. One part is reacted with sulfuric acid and a white precipitate C is formed. Solution A is also reacted with aqueous silver nitrate and a precipitate, D, is formed.

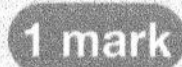

i Write an equation to show the formation of A and B. Identify which is which. 2 marks

ii Write an ionic equation for the formation of B. 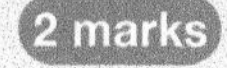 1 mark

iii What is the formula of compound C? 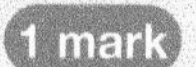 1 mark

iv Identify the precipitate D and state the colour of the precipitate.  2 marks

2 A student carries out an experiment with an aqueous solution of calcium bromide. The student adds aqueous silver nitrate and a precipitation reaction occurs. (8)

a i State what the student observes. 1 mark

ii Write an ionic equation for the reaction that occurs. 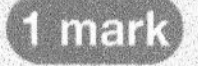 1 mark

b The student is given a second solution of calcium bromide contaminated with sodium iodide. The student carries out two tests to confirm the presence of the iodide ions.

Test 1: the student carries out the same experiment as described in a, but this time adds a second reagent to show the presence of the iodide ions.

i Identify the second reagent the student would need to complete the test. 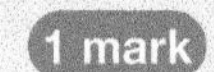 1 mark

....

ii Describe what the student sees when the second reagent is added.  2 marks

....

....

....

....

Test 2: the student also adds aqueous bromine to the contaminated sample of calcium bromide. The student then adds hexane to the mixture and allows it to settle.

iii State, and explain, what the student observes. 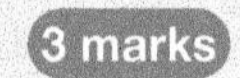 3 marks

....

....

....

....

....

iv Write an ionic equation for the reaction that occurs. 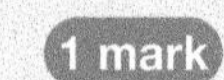 1 mark

....

3 A group of students are asked to design and carry out an experiment to identify a group 2 carbonate (MCO_3). One student decides to react the group 2 carbonate (MCO_3) with hydrochloric acid and collect the carbon dioxide given off. The equation for the reaction is: 12

$$MCO_3(s) + 2HCl(aq) \rightarrow MCl_2(aq) + CO_2(g) + H_2O(l)$$

The apparatus is set up as shown below. Initially the measuring cylinder is filled with water.

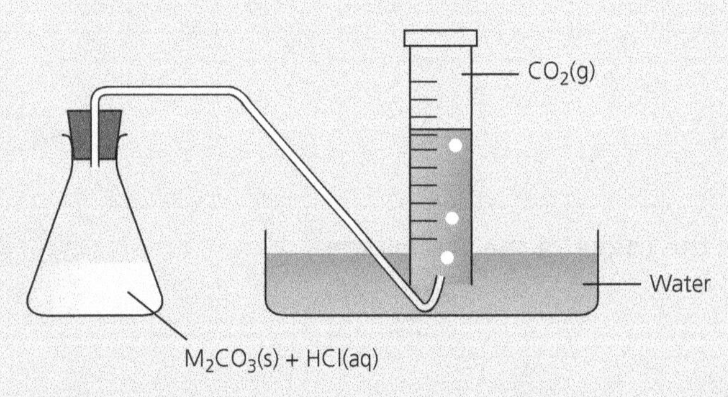

150 cm³ of 1.00 mol dm⁻³ HCl(aq) is added to the conical flask. Approximately 0.70 g of the carbonate is accurately weighed and recorded in the table below. The weighed sample of the carbonate is added to the conical flask and the bung is quickly replaced. The $CO_2(g)$ evolved is collected by the displacement of the water in the measuring cylinder.

The volume of $CO_2(g)$ collected is recorded in the table below.

Mass of the MCO_3 used/g	0.71 g	±0.01
Volume of $CO_2(g)$ collected/cm³	195 cm³	±2.0

a i Calculate the amount in moles of $CO_2(g)$ collected. 1 mark

ii Deduce the amount in moles of $MCO_3(s)$ used.

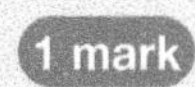

iii Calculate the molar mass of the carbonate. State the units.

iv Identify the group 2 carbonate. Show your working. 2 marks

A second student decides to identify the group 2 carbonate by carrying out a titration. The student accurately weighs 5.10 g of $MCO_3(s)$ and uses a volumetric flask to prepare 250 cm³ of solution. The student then pipettes 25 cm³ of this solution and titrates it with 0.50 mol dm⁻³ HCl(aq).

The student's results are recorded in the table below.

	Trial	1	2	3
Final burette reading/cm³	0.00	0.00	0.00	24.20
Initial burette reading/cm³	25	24.90	24.20	48.50
Titre/cm³	25	24.90	24.20	24.30
Titres used to calculate mean (✓)				
Mean titre/cm³				

b Complete the table by:

i placing a ✓ under the titres that would be used to calculate the mean titre value

ii deducing the mean titre value **2 marks**

c i Write an equation for the reaction between $MCO_3(aq)$ and $HCl(aq)$ **1 mark**

....

ii Calculate the amount in moles of HCl used in the mean titre. **1 mark**

iii Deduce the amount in moles of MCO_3 in $25 cm^3$ of solution. **1 mark**

iv Calculate the amount in moles of MCO_3 in $250 cm^3$ of solution. **1 mark**

v Calculate the molar mass of MCO_3. **1 mark**

vi Identify M. **1 mark**

d Compare the results from each experiment and suggest two sources of error in the less accurate experiment, other than measurement errors. **2 marks**

....

....

....

....

....

....

4 Chlorine is an essential element and many of its compounds are used widely. (12)

a In drinking water, chlorine is used to kill bacteria.

i Write an equation to show the reaction of chlorine with water. **1 mark**

....

ii Some scientists believe that compounds of chlorine should not be added to drinking water. Suggest one reason why some scientists are concerned about the addition of chlorine compounds to drinking water. **1 mark**

....

....

b Chlorine reacts with group 1 elements to form chlorides that are very soluble in water. Aqueous chloride ions can be detected by adding aqueous silver nitrate followed by the addition of aqueous ammonia.

i Write an equation for the reaction between lithium and chlorine. **1 mark**

....

ii Write an ionic equation for the reaction between aqueous silver nitrate and aqueous chloride ions. **1 mark**

....

iii Describe what the student would see when aqueous silver nitrate followed by aqueous ammonia is added to the aqueous chloride ions. **2 marks**

....

....

c A student is given a sample of an unknown group 1 chloride. The student dissolves 6.05 g of the chloride in water and adds excess aqueous silver nitrate. 7.17 g of solid silver chloride is formed.

Identify the group 1 metal. **4 marks**

d Ammonium chloride is a salt that has covalent bonds, dative covalent bonds and ionic bonds. Draw a dot-and-cross diagram of ammonium chloride and clearly label each type of bond. **4 marks**

e A teacher heats 0.613 g of solid $KClO_3$. The equation for the reaction is given below:

$$2KClO_3(s) \rightarrow 2KClO(s) + 3O_2(g)$$

Calculate the volume of oxygen produced (in cm^3) when measured at room temperature and pressure. **4 marks**

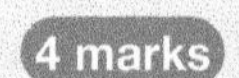

Philip Allan, an imprint of Hodder Education, an Hachette UK company, Blenheim Court, George Street, Banbury, Oxfordshire OX16 5BH

Orders

Hachette UK Distribution, Hely Hutchinson Centre, Milton Road, Didcot, Oxfordshire, OX11 7HH

tel: 01235 827827

e-mail: education@hachette.co.uk

Lines are open 9.00 a.m.–5.00 p.m., Monday to Friday. You can also order through www.hoddereducation.co.uk

ISBN 978-1-4718-4733-2

First printed 2015

Impression number 5

Year 2021

Cover photo is reproduced by permission of Fotolia

Typeset by Aptara, Inc.

Printed in the UK

Hachette UK's policy is to use papers that are natural, renewable and recyclable products and made from wood grown in well-managed forests and other controlled sources. The logging and manufacturing processes are expected to conform to the environmental regulations of the country of origin.